T0093388

Cambridge Elements

Elements in Geochemical Tracers in Earth System Science

NITROGEN ISOTOPES IN DEEP TIME

Colin Mettam
University College London
Aubrey L. Zerkle
University of St Andrews

CAMBRIDGE
UNIVERSITY PRESS

University Printing House, Cambridge CB2 8BS, United Kingdom

One Liberty Plaza, 20th Floor, New York, NY 10006, USA

477 Williamstown Road, Port Melbourne, VIC 3207, Australia

314–321, 3rd Floor, Plot 3, Splendor Forum, Jasola District Centre,
New Delhi – 110025, India

79 Anson Road, #06–04/06, Singapore 079906

Cambridge University Press is part of the University of Cambridge.

It furthers the University's mission by disseminating knowledge in the pursuit of education, learning, and research at the highest international levels of excellence.

www.cambridge.org
Information on this title: www.cambridge.org/9781108810708
DOI: 10.1017/9781108847186

First published 2021

A catalogue record for this publication is available from the British Library.

ISBN 978-1-108-81070-8 Paperback
ISSN 2515-7027 (online)
ISSN 2515-6454 (print)

Nitrogen Isotopes in Deep Time

Elements in Geochemical Tracers in Earth System Science

DOI: 10.1017/9781108847186
First published online: January 2021

Colin Mettam
University College London

Aubrey L. Zerkle
University of St Andrews

Author for correspondence: Aubrey L. Zerkle, az29@st-andrews.ac.uk

Abstract: Nitrogen is an essential nutrient for life, and its sources and cycling have varied over Earth history. Stable isotope ratios of nitrogen compounds (expressed as $\delta^{15}N$, in ‰) are preserved in the sedimentary record and track these changes, providing important insights into associated biogeochemical feedbacks. Here we review the use of nitrogen stable isotope geochemistry in unravelling the evolution of the global N cycle in deep time. We highlight difficulties with preservation, unambiguous interpretations, and local versus global effects. We end with several case studies illustrating how depositional and stratigraphic context is crucial in reliably interpreting $\delta^{15}N$ records in ancient marine sediments, both in ancient anoxic (Archean) and more recent well-oxygenated (Phanerozoic) environments.

Keywords: Nitrogen isotopes, Nutrients, Precambrian Earth, Early life

ISBNs: 9781108810708 (PB), 9781108847186 (OC)
ISSNs: 2515-7027 (online), 2515-6454 (print)

Contents

1 Introduction – Nitrogen in Life

Nitrogen (N) is an essential nutrient for life, as it is critical in the formation of biomolecules, including nucleic acids and proteins. In spite of an abundant reservoir of N_2 in the atmosphere, only a limited number of microorganisms (nitrogen-fixing organisms, or "diazotrophs") have evolved the ability to directly assimilate di-nitrogen. All other organisms rely either on ammonium (NH_4^+) or ammonia (NH_3) released during the remineralization of biomass, or on the products of oxidized ammonium/ammonia generated by biologically mediated transformations in the nitrogen cycle, such as nitrate (NO_3^-) and nitrite (NO_2^-).

Deciphering the evolution of the nitrogen cycle through geological time and the relative abundances of different bioavailable nitrogen compounds is therefore critical to understanding the emergence and radiation of early life, and to elucidating key biological and environmental transitions in the Phanerozoic. Such changes can be tracked by measuring the two stable isotopes of nitrogen in Earth materials (^{14}N and ^{15}N – expressed as $\delta^{15}N$ in ‰, and described below), since different nitrogen cycling reactions express different isotopic fractionations in $\delta^{15}N$. The resulting N isotope ratios are reflected in biomass, which, in turn, can be archived in the sedimentary record (e.g., as reviewed in Stüeken et al., 2016) (Figure 1).

A seminal paper analyzing changes in the marine nitrogen cycle through the Precambrian from temporal $\delta^{15}N$ trends was published by Beaumont and Robert (Beaumont and Robert, 1999). These authors found a shift from $\delta^{15}N$ values in kerogen centred around 0‰ in the Archaean, to positive $\delta^{15}N$ values centred around +5‰ from the Paleoproterozoic onwards. They suggested that these records were broadly indicative of an anaerobic NH_4^+-based nitrogen cycle in the Archean, giving way to an aerobic nitrogen cycle with available nitrite and nitrate after the Great Oxidation Event (GOE). In the ensuing two decades, the $\delta^{15}N$ proxy has become increasingly utilized for paleoenvironmental and paleoredox studies in deep time (here defined as pre-Cenozoic; e.g., Ader et al., 2016; Stüeken et al., 2016a), with more recent targeted studies largely supporting these broad temporal trends (e.g., Kipp et al., 2018; Yang et al., 2019; Zerkle et al., 2017).

The general stepwise expansion of aerobic nitrogen cycling notwithstanding, nitrogen isotope studies of sediments spanning Earth's history have revealed spatial and temporal nuances within this narrative. For example, some $\delta^{15}N$ trends in Late Archean sediments have been interpreted to represent periods of temporally and spatially constrained aerobic N cycling prior to the GOE (Garvin et al., 2009; Godfrey and Falkowski, 2009; Mettam et al., 2019; Yang et al., 2019). More recent $\delta^{15}N$ records have identified periods of time when anaerobic nitrogen cycling dominated in the Phanerozoic, including during the Latest Permian Extinction

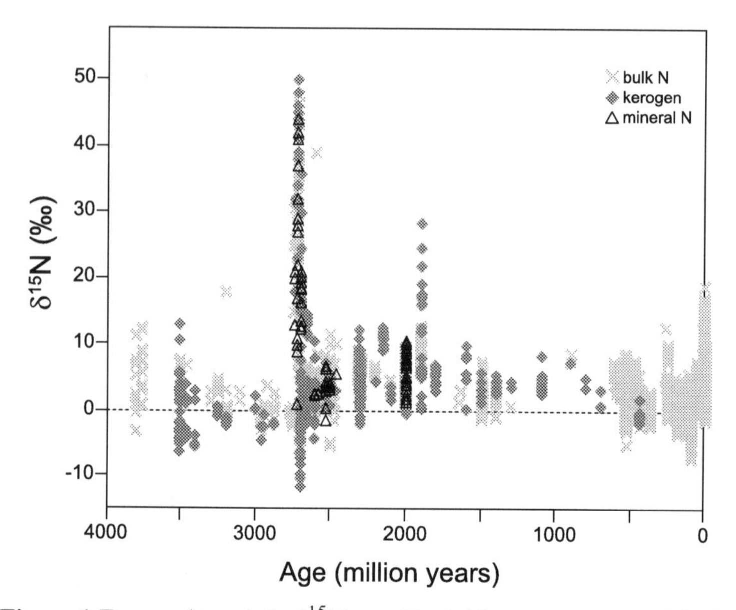

Figure 1 Temporal trends in $\delta^{15}N$ over Earth history, as preserved in the sedimentary rock record (updated from Yang et al., 2019, and references therein)

Event (LPEE; e.g., Saitoh et al., 2014) and Cretaceous Ocean Anoxic Events (OAEs; e.g., Junium and Arthur, 2007).

Interpretations of these records, and the biogeochemical feedbacks they imply, critically rely on the ability of sedimentary $\delta^{15}N$ to reliably record the isotopic fingerprints of the coeval marine nitrogen cycle. Here we discuss the state of the field in low-temperature N isotope biogeochemistry, including susceptibility to syn- and post-depositional alteration, ambiguities with respect to interpretations, and local versus global effects. Notably, we argue that stratigraphic and depositional context is crucial in reliably interpreting nitrogen stable isotope records in ancient marine sediments.

2 The $\delta^{15}N$ Proxy in a Nutshell

Multiple biologically mediated redox transitions occur between nitrogen-containing compounds in the marine N cycle. These transitions and their associated $\delta^{15}N$ fractionations are summarized in Figure 2. The $\delta^{15}N$ values of nitrogen compounds are expressed relative to the isotopic composition of atmospheric nitrogen, using the standard delta notation:

$$\delta^{15}N\ (‰) = ((^{15}N/^{14}N)_{sample} / (^{15}N/^{14}N)_{air} - 1) \times 1000,$$

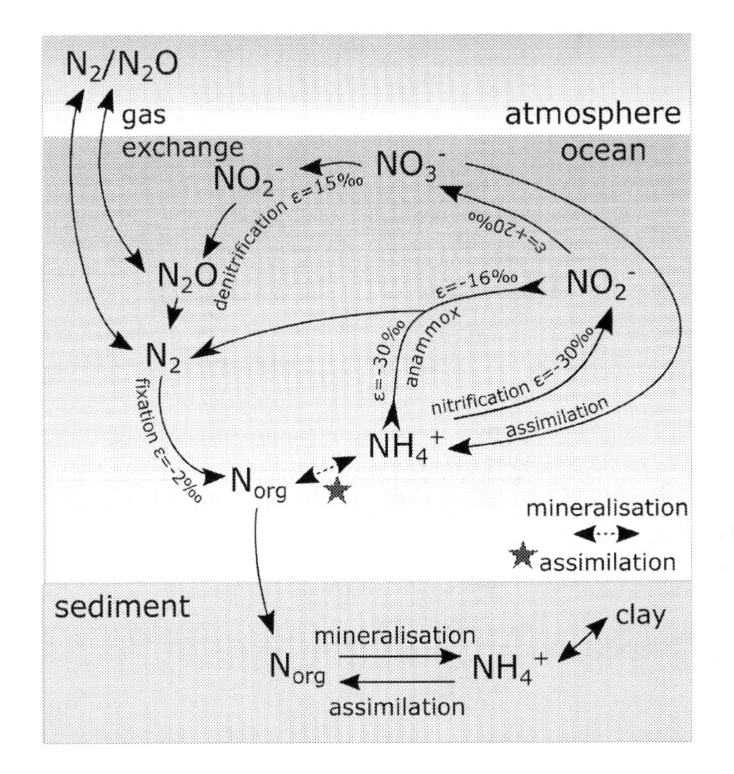

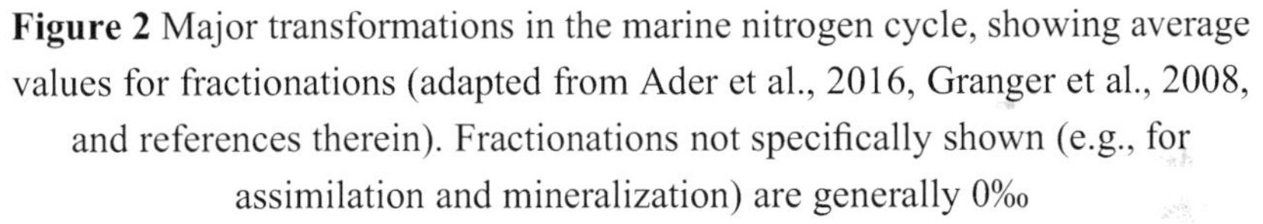
Figure 2 Major transformations in the marine nitrogen cycle, showing average values for fractionations (adapted from Ader et al., 2016, Granger et al., 2008, and references therein). Fractionations not specifically shown (e.g., for assimilation and mineralization) are generally 0‰

with isotopic fractionations for specific reactions shown as $\varepsilon \approx \delta^{15}N_{product} - \delta^{15}N_{reactant}$.

N_2 fixation is the primary source of nitrogen input into the marine system and is regulated by the availability of phosphorous and micronutrients, such as Mo and Fe (e.g., Zerkle et al., 2006). Diazotrophs incorporate nitrogen into their biomass directly from atmospheric or dissolved N_2. Remineralization of this biomass in the water column and in sediments releases bioavailable ammonium or ammonia, which can, in turn, be assimilated by non-diazotrophic organisms. N_2 fixation and the recycling of NH_4^+ generally impart only small fractionations in $\delta^{15}N$ ($\varepsilon \leq 2‰$). Therefore, in an environment in which N_2 fixation and the recycling of NH_4^+ are the dominant N sources, the $\delta^{15}N$ of biomass broadly reflects the atmospheric input value ($\delta^{15}N \approx 0‰ \pm 2‰$). However, if NH_4^+ or other forms of dissolved inorganic nitrogen (DIN) are readily available, diazotrophy will be suppressed, as N_2 fixation is energetically expensive in comparison to assimilation.

In the presence of oxygen, ammonium/ammonia undergoes microbially mediated sequential oxidation to nitrite (NO_2^-) and nitrate (NO_3^-). In modern oxygen-rich settings, this biological nitrification process is rapid and quantitative, making NO_3^- the largest reservoir of bioavailable DIN in the oceans. Although nitrification can produce large fractionations in $\delta^{15}N$, the quantitative nature of NH_4^+ oxidation in the modern oceans means that these isotopic fractionations are suppressed in the resultant nitrate pool.

Nitrate and nitrite both form important sources of nutrient N in modern oceanic settings; however, they can also be utilized as electron acceptors in chemotrophic metabolisms. Nitrate can be reduced to N_2 during heterotrophic denitrification, a form of anaerobic respiration of organic carbon with NO_3^-, which is second only to aerobic respiration in reduction potential. This canonical form of denitrification proceeds through NO_2^- and a number of intermediate N phases that can also build up in the environment (e.g., N_2O). Nitrate reduction can further be coupled to the oxidation of reduced compounds such as sulfide, methane, or hydrogen during chemoautotrophy. Nitrate reduction can also proceed via dissimilatory reduction to ammonium (DNRA), which provides a competitive advantage under nitrate-limiting conditions since it requires less nitrate per mole of organic substrate. Finally, some organisms can utilize nitrite to oxidize NH_4^+ during anaerobic ammonium oxidation (anammox). The reduction of NO_3^-/NO_2^- to N_2 during denitrification and anammox can have important implications for the oceanic nitrogen budget, as these processes remove bioavailable nitrogen from the oceanic reservoir. In the case of DIN loss, and if other nutrients remain available, diazotrophs will have a competitive advantage, and can proliferate to restore the balance of fixed N to the marine system (e.g., Megonigal et al., 2003; Voss et al., 2012).

These N loss processes can all produce large fractionations in $\delta^{15}N$ ($\varepsilon = +20$ to $+30‰$; Brunner et al., 2013; Granger et al., 2008). In marine sediments, denitrification is generally quantitative such that no fractionations are expressed. However, in the water column, denitrification is constrained to low-oxygen settings such as oxygen minimum zones (OMZs; Figure 3). This distribution results in incomplete denitrification in the water column, leaving a residual pool of NO_3^- that carries a positive $\delta^{15}N$ value. Anammox produces a similar fractionation effect for $\delta^{15}N$, but its role in the global N cycle and in the N isotope budget is somewhat less well-constrained and the relative contributions of anammox and canonical denitrification remain an area of active research.

The balance between N_2 fixation and the assimilation of NO_3^- carrying the positive $\delta^{15}N$ signature of denitrification/anammox controls the $\delta^{15}N$ values of modern marine biomass. Today, aerobic N loss processes produce organic matter with an average $\delta^{15}N$ of ~+6‰ (Peters et al., 1978). However, the $\delta^{15}N$ of particulate organic matter and underlying sediments is spatially and

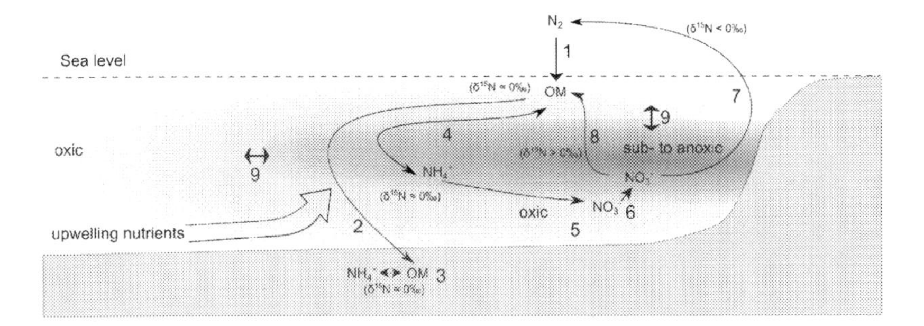

Figure 3 Nitrogen cycling in a Phanerozoic Oxygen Minimum Zone (OMZ) (adapted from Pinti and Hashizume, 2011). (1) Fixation of gaseous nitrogen (N_2) into biomass (organic matter, OM) by diazotrophs; (2) transport of OM to the sediments, i.e., via the biologic pump; (3) OM can be preserved in the sediments or remineralized to NH_4^+, which can then be reassimilated into OM by benthic organisms or adsorbed onto clay minerals; (4) remineralization of OM to NH_4^+ and reassimilation of NH_4^+ in low-oxygen settings; (5) in oxic settings NH_4^+ is oxidized to NO_2^- and NO_3^-; (6) upwelling of NO_3^- into low-oxygen setting; (7) partial denitrification of NO_3^- (and reduction of NO_2^- by anammox) in sub- to anoxic settings releases N_2 to atmosphere. Partial denitrification renders the residual pool of NO_3^- isotopically heavy (+ $\delta^{15}N$); (8) assimilation of this residual pool of NO_3^- contributes to the positive $\delta^{15}N$ values of Phanerozoic OM; (9) the size of OMZs can expand and contract enhancing or reducing denitrification and the resulting N isotope effects

temporally heterogeneous, following subtle redox and productivity variations in the modern marine system (e.g., De Pol-Holz et al., 2009). In addition, changes in the redox state and nutrient dynamics of marine settings over Earth's history have seemingly allowed for unique configurations of the nitrogen cycle that produce both highly positive and highly negative $\delta^{15}N$ values not seen on Earth today (Figure 1). We explore some of these records, and their potential (largely non-unique) interpretations in the case studies below.

3 Methods and Limitations

Nitrogen isotopes are now routinely measured alongside carbon isotopes in sedimentary rocks. Most recent studies have measured sedimentary $\delta^{15}N$ using an elemental analyzer coupled to an isotope ratio mass spectrometer running in continuous flow mode (CF-EA-IRMS). This method is more widely accessible and significantly less time consuming than previously developed offline methods (see Ader et al., 2016, for a comprehensive review). The CF-EA-IRMS method generally works well for sediments with high N contents (1σ of

±0.25‰ for N > 700ppm; Bahlmann et al., 2010). However, the low N content and high TOC/TN ratios in most Precambrian rocks can lead to some analytical challenges. In this case, additional steps are required to correct for blanks and ensure complete combustion to prevent isobaric interferences, e.g., with CO (Beaumont et al., 1994).

Several new methods have been developed in the past decade that also allow for accurate $\delta^{15}N$ measurements on samples with much lower abundances of N. The nano-EA-IRMS method (Polissar et al., 2009) uses a custom cryo-trap and chromatography system to concentrate the evolved N_2 gas before analysis, decreasing the sample size requirements by several orders of magnitude. This method has been successfully applied to analyze $\delta^{15}N$ values in deep time sediments (e.g., Mettam et al., 2019; Yang et al., 2019), but blanks remain an issue and must be closely monitored and corrected for. Alternatively, samples can be run in triplicates of varying mass. These triplicates can then be plotted on "Keeling" plots, with the true value calculated as the intercept with the y-axis (see Mettam et al., 2019). An offline tube cracker combustion system has also been successfully applied as an introduction system to the CF-EA-IRMS to measure $\delta^{15}N$ in low N abundance sediments (e.g., Stüeken et al., 2015a), but requires very high vacuum. Ishida et al. (2018) recently described a method for measuring $\delta^{15}N$ values in organic matter in situ using secondary ion mass spectrometry (SIMS), with analytical precision on the same order as bulk methods (2σ of ±0.56‰). Spatially resolved $\delta^{15}N$ analyses should be useful in discerning $\delta^{15}N$ at the single cell or organismal level (e.g., with Precambrian microfossils); however, the SIMS technique measures $\delta^{15}N$ on the CN^- ion, so cannot be used to measure mineral NH_4^+, e.g., in clay minerals that don't contain carbon.

Despite a recent upswing in the use of $\delta^{15}N$ values in deep time paleoenvironmental and paleoredox studies, post-depositional alteration remains an issue, and the relative ability of different sedimentary N phases to resist alteration is hotly debated. In the following sections, we discuss post-depositional alteration of N stable isotope ratios, and the consensus (or lack thereof) for which N phases most faithfully retain their original $\delta^{15}N$ signatures.

3.1 The Problem with Preservation

In order to reconstruct ancient nitrogen cycling from the sedimentary rock record, the organic matter preserved within these sediments must faithfully record the $\delta^{15}N$ values of the coeval marine biomass. Post-mortem, marine phytoplankton will sink to the sea floor, aided by the biological pump (at least in the Phanerozoic). Some of this organic matter will be remineralized in the water column or directly consumed on descent, which could affect $\delta^{15}N$ values. Modern N isotope studies have shown that the oxidation of sinking particulate matter can be an important consideration for preservation, as deamination of organic matter can selectively liberate ^{15}N-depleted ammonium, thereby slightly increasing the residual $\delta^{15}N$ values (Möbius et al., 2010).

Primary N isotope values can be further modified by syn- and post-depositional processes in the sediments, including diagenesis, metamorphism, and hydrothermal or hydrocarbon fluid migration. Numerous theoretical and empirical studies have been utilized to estimate these effects (e.g., see review in Ader et al., 2016). While the detailed findings of these studies differ, the processes that seem to impart the largest changes in the primary $\delta^{15}N$ values include aerobic degradation of organic matter during early diagenesis, which can increase $\delta^{15}N$ values by 2–4‰ (Freudenthal et al., 2001), and high-temperature metamorphism above greenschist facies, which can increase $\delta^{15}N$ values by up to 5‰ (Bebout and Fogel, 1992; Boyd and Philippot, 1998).

Nitrogen isotope values can be modified by syn-depositional degradation of organic matter within the sediments, either by aerobic or anaerobic respiration (e.g., sulphate reduction) (Altabet and Francois, 1994; Macko and Estep, 1984). However, these ^{15}N enrichments can be muted if the evolved NH_4^+ is quantitatively reincorporated into new benthic biomass or adsorbed onto clay mineral phases (Robinson et al., 2012). Potential loss of nitrogen during degradation can be qualitatively assessed by examining the ratios of organic carbon to total nitrogen (C/N ratios) in the sediments. The Redfield ratio for pristine marine phytoplankton is between ~4 and ~8 (Gao et al., 2012; Meyers, 1997). Sedimentary values below this lower threshold would indicate that a significant amount of carbon was lost during degradation, while liberated NH_4^+ was largely captured and retained on clay surfaces. These considerations are complicated for many Precambrian sediments, where C/N ratios are extremely high (> 100). In addition, the use of C/N ratios to infer preservation in Phanerozoic sediments can be complicated by mixing with terrestrial vegetation, which has a higher C/N ratio of > ~15 (Gao et al., 2012; Meyers, 1997); such mixed provenance can potentially be untangled by coeval biomarker analyses, as discussed below.

Modification of sedimentary $\delta^{15}N$ values by metamorphism is particularly well-documented, and generally increases with metamorphic grade. Thermal alteration results in the preferential liberation of the lighter stable isotope, which increases $\delta^{15}N$ and $\delta^{13}C$ values. These effects can be identified by a diagnostic positive correlation between $\delta^{15}N$ and $\delta^{13}C$, and by negative correlations between total nitrogen (TN) and $\delta^{15}N$, or organic carbon (TOC) and $\delta^{13}C$. Stüeken et al. (2017) further proposed that C/H ratios in kerogen can be used as a proxy for thermal alteration in metamorphosed sediments. Regardless, metamorphic effects on $\delta^{15}N$ are generally mild (1–2‰) at greenschist facies or below, thus the majority of sedimentary $\delta^{15}N$ studies have focused on low metamorphic grade sediments.

3.2 The $\delta^{15}N_{org}$ versus $\delta^{15}N_{bulk}$ Debate

Issues surrounding post-burial preservation and alteration are also intrinsic to the debate over whether $\delta^{15}N$ values in bulk sediments ($\delta^{15}N_{bulk}$) or in extracted organics ($\delta^{15}N_{org}$) are the more reliable recorder of sedimentary organic nitrogen. The former includes a combination of nitrogen contained in organic matter and NH_4^+ adsorbed onto clay minerals or substituted into minerals as a replacement for potassium (K^+), while the latter represents the acid-resistant organic fraction. This debate stems from discrepancies between $\delta^{15}N_{bulk}$ and $\delta^{15}N_{org}$ values, which often show an offset between isotopically lighter $\delta^{15}N_{org}$ and heavier $\delta^{15}N_{bulk}$ within the same sample.

Proponents of analyzing sedimentary organic matter (or kerogen) argue that this archive is less easily altered or exchangeable than more mobile N pools, such as clay-adsorbed NH_4^+. However, concerns remain about the efficiency and selectivity of specific methods for kerogen extraction. Most kerogen extraction techniques use a series of rock powder dissolutions with increasingly corrosive acids to dissolve the carbonate and silicate fractions (often HCl-HF). Notably, recalcitrant minerals that are also resistant to HF digestion, such as pyrite, can be retained through these extraction procedures. While these minerals do not generally contain N, they do influence the nitrogen abundance values of kerogen (total organic nitrogen, TON wt. %), complicating interpretations of TON. These minerals can be removed by the inclusion of a heavy mineral separation step after acid digestion, e.g., with chloroform. A greater concern is the formation of neo-minerals, such as fluoride salts, which can alter measured $\delta^{15}N$ values during the combustion of samples in EA-IRMS. The formation of calcium fluoride (CaF) can be prevented to some extent by digestion of sediments in HCl and thorough rinsing to remove carbonate-associated Ca^{2+} prior to HF extraction; however, this remains an issue for samples with high abundances of Ca-bearing silicates. An additional digestion using a combination of HF and boric acid to produce BF_3 has recently proved effective in removing secondary fluorides (Stüeken et al., 2017).

Proponents of analyzing bulk sediments suggest that the offset between $\delta^{15}N_{org}$ and $\delta^{15}N_{bulk}$ from the same sample indicates that significant amounts of ^{15}N must have been liberated from kerogen and ultimately captured on clays. Recent work by Stüeken et al. (2017) suggested that metamorphism drives organic $\delta^{15}N$ values lighter and silicate-bound $\delta^{15}N$ values heavier, resulting in a 3–4‰ offset at greenschist facies or lower. However, the mechanism(s) behind this offset remain unclear. If this difference occurs when NH_4^+ is transferred from kerogen to clays during early diagenesis, it could result from the preferential removal of isotopically heavy protein-derived organic matter from kerogen or the preferential preservation of isotopically light porphyrins in the kerogen; if it occurs during thermal maturation, it could be caused by changes in bonding as kerogen matures and clays dewater (Stüeken et al., 2017).

These arguments imply that $\delta^{15}N_{bulk}$ could provide the more reliable sedimentary archive, particularly when the majority of nitrogen resides in silicates phases. However, the retention of primary $\delta^{15}N$ values in bulk sediments requires that post-depositional processes occur within a closed system that captures most of the NH_4^+ evolved from organic matter (e.g., Robinson et al., 2012). In addition, $\delta^{15}N_{bulk}$ data are susceptible to overprinting and isotope exchange between clay-bound NH_4^+ and NH_4^+ from migrating metasomatic fluids. Kump et al. (2011) suggested that potential overwriting by metasomatic NH_4^+ should be identifiable by strong correlations between TN, $\delta^{15}N_{bulk}$, and potassium (K, wt%), given the abundance of K in metasomatic fluids. Similar correlations should also identify contamination by allochthonous clay-associated NH_4^+ brought to the depocentre from eroding hinterlands, although such contamination with exogenous N is less easily identified. Given the propensity for these post-depositional processes to alter different N phases, and the lack of a consensus from the community as to which record is more reliable, many studies now report both $\delta^{15}N_{bulk}$ and $\delta^{15}N_{org}$ data, and comparison of these two datasets can reveal further insights into data fidelity (e.g., Koehler et al., 2018; Yang et al., 2019).

4 Case Studies

Taking into account the precautions discussed earlier, $\delta^{15}N$ values provide important insights into past biogeochemical nitrogen cycling in well-preserved sedimentary rocks. Reliable $\delta^{15}N$ data have been generated in sediments as old as 3.8 Ga (Papineau et al., 2005), and provide an interesting narrative of marine nutrient cycling following from biological revolutions that occurred over Earth history. Even the earliest $\delta^{15}N$ records show evidence for N_2 fixation (Stüeken et al., 2015a) consistent with phylogenetic suggestions that diazotrophy arose very early in life's history (Weiss et al., 2016). As discussed already, the long-term narrative of nitrogen cycling through geological time documents a change from dominantly anaerobic to aerobic N cycling, consistent with the progressive oxygenation of the biosphere. However, recently published records hint at earlier spatial and temporal heterogeneities, even in a largely anoxic Late Archean world. In addition, records of $\delta^{15}N$ are increasingly being utilized to examine the role of nutrients in more recent transitions in Earth history, including mass extinction events in the Phanerozoic. Here, we present three case studies illustrating the utility and complexity of the $\delta^{15}N$ proxy in deciphering the biogeochemical nitrogen cycling in the Late Archean and Late Permian marine systems.

4.1 Extreme $\delta^{15}N$ Values at ~2.7 Ga – Marine versus Terrestrial Signals?

Both extremely high and extremely low $\delta^{15}N$ values have been measured in sediments from ~2.7 Ga (Figure 1), and variably linked to global oxygenation and primary production. $\delta^{15}N$ values as high as +50‰ in the Tumbiana Formation (Fortescue Group, Western Australia) have been interpreted to represent partial nitrification under conditions in which ammonium was abundant and oxygen was limited (Thomazo et al., 2011). This interpretation stems largely from laboratory incubations with methanotrophic bacteria that produced extremely ^{15}N-depleted N_2O (–46‰) during the co-oxidization of ammonium with methane (Mandernack et al., 2009). The co-occurrence of these large $\delta^{15}N$ values with very low $\delta^{13}C_{org}$ values in the Tumbiana Formation supports the incorporation of methane into these sediments. However, this interpretation remains problematic for several reasons (e.g., Stüeken et al., 2015b). Namely, in order for isotopic fractionations this large to be expressed in the residual NH_4^+, a large fraction of the available ammonium (> 70%) would have had to be oxidized. The residual ammonium would still have been assimilated into biomass, while the resulting NO_2^- and NO_3^- would have been respired. Similarly, in order for the high $\delta^{15}N$ values to have been preserved in the resulting organic matter, the product NO_2^- would have to be lost from the system (e.g., via denitrification and/or anammox) rather than reassimilated. Stüeken et al. (2015b) alternatively suggested that the Tumbiana Formation and underlying Kylena Formation represent a lake system that was driven highly alkaline by aqueous alteration of underlying volcanics. In an anoxic alkaline lake setting, the volatilization of NH_4^+ to ammonia (NH_3) could produce large N isotope fractionations, similar to high $\delta^{15}N$ values in alkaline lakes today.

Extremely low $\delta^{15}N$ values have recently been reported from ~2.7 Ga sediments of the Manjeri Formation (Belingwe Greenstone Belt, Zimbabwe) (Yang et al., 2019). These values, down to –11‰, were interpreted to form from incomplete assimilation of a non-limiting pool of upwelling NH_4^+. This scenario was linked to a rise in global primary productivity following the expansion of oxygenic photosynthesis. The associated surge in primary productivity could have enhanced organic matter export and remineralization, leading to the accumulation of NH_4^+ in anoxic deep waters. This pool of NH_4^+ could have been upwelled into highly productive surface oceans and partially assimilated by primary producers preferentially utilizing isotopically light NH_4^+. In addition, uptake of light NH_4^+ would have left the residual pool of DIN isotopically heavy. Assimilation of this residual pool of isotopically heavy NH_4^+, separated either in time or space, could also have contributed to production of positive $\delta^{15}N$ values in the Late Archean (Ader et al., 2016; Yang et al., 2019).

4.2 Pre-GOE Positive $\delta^{15}N$ Values – Alternatives to Oxygen?

In spite of the general narrative of an anaerobic Archean N cycle giving way to aerobic N cycling sometime later in Earth history, the exact timing of this transition is debated. Some statistical treatments of the secular $\delta^{15}N$ record seem to imply that nitrate was widely available by ~2.5 Ga (Stüeken et al., 2016), while others suggest that a turning point in the N cycle occurred in association with the GOE at ~2.3 Ga (Zerkle et al., 2017). Positive $\delta^{15}N$ values in Late Archean sediments have also been interpreted to imply the early onset of aerobic N cycling; however, these data are not without controversy, and lend themselves to alternative interpretations.

Small increases in $\delta^{15}N$ values preserved within ~2.5 Ga sediments from the Mt McRae Formation in Western Australia (Garvin et al., 2009) and the Ghaap Group in South Africa (Godfrey and Falkowski, 2009) have been suggested to record the transient or localized appearance of nitrification and denitrification in association with "whiffs" of oxygen in the marine environment. However, these sediments were deposited in distal and relatively deep waters, separated from oxygen oases in shallow, highly productive shelf environments in which oxygen production was most likely to have occurred (e.g., Olsen et al., 2013). Therefore, these sediments could instead represent nitrogen redox cycling in open ocean or deeper water environments independent of surface oxygen. For example, positive $\delta^{15}N$ values preserved in ~2.5 Ga BIFs have been interpreted to reflect the assimilation of NH_4^+ enriched in ^{15}N by partial oxidation to nitrite. However, these researchers argued that ammonium oxidation could have been driven anaerobically by microbes utilizing Fe(III) oxyhydroxides formed in the water column, rather than by O_2 (Busigny et al., 2013).

$\delta^{15}N$ values of ammonium in the Late Archean could also have been driven higher by partial assimilation, as described for ~2.7 Ga sediments earlier (Ader et al., 2016). Until recently, this interpretation was largely dismissed due to a lack of evidence for a correlative pool of ^{15}N-depleted biomass. However, recent analyses of ~2.5 Ga rocks have identified comparatively low $\delta^{15}N$ values (≥ -4‰) in shore-proximal sediments (Mettam et al., 2019). Combined with the ~2.7 Ga data (Yang et al., 2019), these studies suggest that partial NH_4^+ assimilation could have been widespread once oxygenic photosynthesis kickstarted primary productivity in the Late Archean. These data additionally illustrate the potential for spatial and temporal heterogeneity within the global marine N cycle.

The study by Mettam et al. (2019) further demonstrates how depositional setting can play a key role in spatially partitioning the nitrogen cycle, producing local variations in $\delta^{15}N$ values. For example, $\delta^{15}N$ values of –4‰ were measured in sediments deposited from relatively deep shelf conditions, where partial assimilation from a pool of upwelling NH_4^+ from the open ocean could have occurred. In contrast, sediments deposited in shallower, more restricted lagoonal conditions retained $\delta^{15}N$ values of ~0‰, likely reflecting diazotrophy and the efficient recycling of remineralized NH_4^+. Only a handful of positive $\delta^{15}N$ values greater than 2‰ were reported from the same section, hinting at the possible presence of coupled nitrification and incomplete denitrification (Figure 4). However, these positive values were from shallow-water, carbonate-rich facies, probably indicating very localized oxygenated settings. This spatial complexity highlights the importance of sedimentological and depositional context for understanding local variations in $\delta^{15}N$ data.

4.3 The Late Permian Extinction Event – Multiple Controls on $\delta^{15}N$?

Trends in Phanerozoic $\delta^{15}N$ values have also been utilized to examine changes in redox and nutrient feedbacks during more recent events in Earth history. For example, nutrient stress during the Late Permian has been implicated in contributing to the greatest extinction event in the Phanerozoic. Notably, $\delta^{15}N$ values near 0‰ in palaeoequatorial Tethyan seas during the LPEE have been interpreted to reflect enhanced denitrification and a proliferation of diazotrophy resulting from NO_3^- limitation (as reviewed by Saitoh et al., 2014).

Decreases of ~1‰ in $\delta^{15}N$ values in late Permian Boreal and Panthalassic waters have also been reported (Algeo et al., 2012; Grasby et al., 2015; Knies et al., 2013; Schoepfer et al., 2012). However, with the exception of one study (Schoepfer et al., 2012), these changes are small and $\delta^{15}N$ values never fall below +4‰. The persistence of positive $\delta^{15}N$ values indicates that these waters probably maintained a robust NO_3^- inventory, which would have precluded an expansion in diazotrophy.

Small decreases in $\delta^{15}N$ values in LPEE sediments could be caused by several other factors. As described earlier, differing rates of organic matter

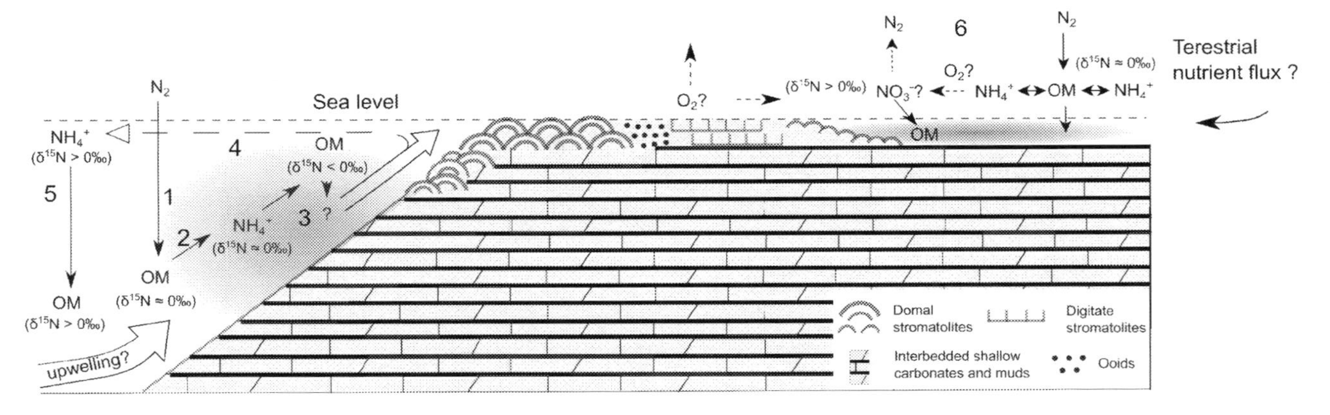

Figure 4 Proposed nitrogen cycle in the Late Archean (adapted from Mettam et al., 2019). Nitrogen cycling was likely spatially heterogenous, as follows: (1) N_2 fixation in the open ocean and transport of diazotrophic biomass to the seafloor; (2) remineralization of OM to NH_4^+ and shoreward transport; (3) incomplete NH_4^+ assimilation, producing OM with negative $\delta^{15}N$ values; (4) transport of the residual NH_4^+ pool (left with positive $\delta^{15}N$ values); (5) complete assimilation of NH_4^+ in the open ocean, producing OM with positive $\delta^{15}N$ values. (6) In a relatively restricted lagoonal environment isolated from marine influence, these processes could alternatively go to completion, such that the $\delta^{15}N$ of OM would reflect the input (0‰). Alternatively, if oxygen oases are present, coupled nitrification/dentrification could produce OM with positive $\delta^{15}N$ values

degradation tied to redox changes can influence the $\delta^{15}N$ values of organic matter during deposition, particularly where fluxes of organic matter are low (Freudenthal et al., 2001; Möbius et al., 2010). Such effects are illustrated by LPEE data from Schuchert Dal, East Greenland (Mettam et al., 2017). C/N ratios in these sediments never exceed 6, indicating little variation in organic matter provenance (e.g., due to terrestrial input). However, higher $\delta^{15}N$ values and lower C/N ratios correspond with oxic, bioturbated depositional horizons, while lower $\delta^{15}N$ values and higher C/N ratios are recorded in oxygen- deficient, laminated horizons. Given these observations, a wholesale transition from oxic to anoxic conditions during the LPEE coupled with slower syn-depositional degradation could also contribute to decreases in $\delta^{15}N$ values at some sites during and after the LPEE.

The complexity of interpreting sedimentary nitrogen isotope values is further highlighted by $\delta^{15}N$ data from Buchanan Lake, Canada, which was part of the Sverdrup Basin on the northwest margin of Pangea during the LPEE (Knies et al., 2013). These authors concluded that increased diazotrophy was likely responsible for a small decrease in $\delta^{15}N$ (from ~+9‰ to ~+8‰) in these sediments. However, persistently positive $\delta^{15}N$ values throughout this section suggest continued nitrate availability in spite of evidence for an intensification of anoxia. A fall in the rate of organic matter degradation associated with enhanced anoxia could be responsible for the small fall in $\delta^{15}N$; however, this scenario is inconsistent with a coeval decrease in C/N ratios. An alternate explanation could come from increased weathering and transport of exposed soils and terrestrial vegetation, consistent with the reorganization of terrestrial ecosystems during the LPEE (Algeo and Twitchett, 2010; Looy et al., 2001). Increased fluxes of terrestrial organic matter are unlikely to be the cause of the decrease in $\delta^{15}N$, as this would most likely increase C/N ratios as well. Alternatively, increased fluxes of clay-adsorbed inorganic nitrogen from the depositional hinterland could change $\delta^{15}N$ while reducing C/N ratios, as seen at this site. Given these alternative arguments, an increase in marine diazotrophy is a reasonable, although perhaps not completely unique, interpretation.

5 Future Perspectives

These case studies highlight the potential complexity of the sedimentary $\delta^{15}N$ record and some of the uncertainties associated with interpreting $\delta^{15}N$ values in deep time. In particular, the paucity of well-preserved Archean sediments and inherent analytical limitations mean that the majority of $\delta^{15}N$ studies have focused on organic carbon-rich sediments deposited in low-energy, deeper water settings. These studies provide important constraints on local nitrogen cycling processes occurring in these basins. However, the spatial heterogeneity of the marine N cycle, along with the ability of disparate N cycling processes to produce overlapping $\delta^{15}N$ values, suggests that depositional and redox context is critical in interpreting the resulting $\delta^{15}N$ records, and in unravelling local versus global signals. The interpretation of Phanerozoic $\delta^{15}N$ records should similarly be approached with caution given the multiple factors that can modify or overprint the isotopic characteristics of primary marine organic matter. For example, redox variations can also influence organic matter degradation rates in the sediments. Furthermore, the evolution of land plants during the Palaeozoic provide a source of organic matter with distinct $\delta^{15}N$ values and C/N ratios, and the transport of these materials to the sediments, along with nitrogen in eroded soils, can lead to mixing of isotope signals that complicate interpretations. Additional targets for future and ongoing development include:

1. *Analytical advancements.* New methods are being developed to characterize the distribution of $\delta^{15}N$ values in low N abundance sediments and to identify post-depositional alteration. In particular, recent advances in in situ $\delta^{15}N$ analyses via SIMS will allow for direct analyses of $\delta^{15}N$ in organic nitrogen (Ishida et al., 2018), which should allow for single-cell $\delta^{15}N$ analyses in Precambrian microfossils.
2. *Field studies.* Modern anoxic aqueous environments, such as redox-stratified lakes and seas, can provide near analogues to Precambrian marine systems. Thus $\delta^{15}N$ values of aqueous and sedimentary N compounds in these systems can be linked directly to N cycling processes, providing important insights into sedimentary $\delta^{15}N$ values preserved in ancient sediments (e.g., Fulton et al., 2018).
3. *Laboratory studies.* N isotopic fractionations produced during cyanobacterial N_2 fixation and denitrification are generally well-calibrated for modern conditions (Baursachs et al., 2009; Granger et al., 2008), but life and environments have evolved significantly over Earth history. Past conditions could have promoted different N cycling processes, by different groups of organisms and/or the same organisms utilizing different enzymes, but the resulting isotopic fractionations are largely unconstrained (e.g., Nishizawa et al., 2014). Incomplete ammonium assimilation in particular has been implicated in contributing to Late Archean $\delta^{15}N$ records, but our understanding of $\delta^{15}N$ fractionations during this process and their response to changing environmental conditions is extremely limited (Hoch et al., 1992).

Combined, these types of inter-disciplinary studies could provide the forward strides necessary for generating and understanding sedimentary $\delta^{15}N$ records throughout Earth history.

References

[**NB** Key references in bold.]

Ader, M., Thomazo, C., Sansjofre, P., Busigny, V., Papineau, D., Laffont, R., Cartigny, P. , and Halverson, G. P., 2016, Interpretation of the nitrogen isotopic composition of Precambrian sedimentary rocks: Assumptions and perspectives: Chemical Geology, v. 429, pp. 93–110.

Algeo, T., Henderson, C. M., Ellwood, B., Rowe, H., Elswick, E., Bates, S., Lyons, T., Hower, J. C., Smith, C., Maynard, B., Hays, L. E., Summons, R. E., Fulton, J. M., and Freeman, K. H., 2012, Evidence for a diachronous Late Permian marine crisis from the Canadian Arctic region: GSA Bulletin, v. 124, pp. 1424–48.

Algeo, T. J. and Twitchett, R. J., 2010, Anomalous Early Triassic sediment fluxes due to elevated weathering rates and their biological consequences: Geology, v. 38, pp. 1023–6.

Altabet, M. A. and Francois, R., 1994, Sedimentary nitrogen isotopic ratio as a recorder for surface ocean nitrate utilization: Global Biogeochemical Cycles, v. 8, no. 1, pp. 103–16.

Bahlmann, E., Bernasconi, S. M., Bouillon, S., Houtekamer, M., Korntheuer, M., Langenberg, F., Mayr, C., Metzke, M., Middelburg, J. J., Nagel, B., Struck, U., Voss, M., and Emeis, K. C., 2010, Performance evaluation of nitrogen isotope ratio determination in marine and lacustrine sediments: An inter-laboratory comparison: Organic Geochemistry, v. 41, pp. 3–12.

Baursachs, T., Schouten, S., Compaore, J., Wollenzien, U., Stal, L. J., and Damste, J. S. S., 2009, Nitrogen isotopic fractionation associated with growth on dinitrogen gas and nitrate by cyanobacteria: Limnology and Oceanography, v. 54, pp. 1403–11.

Beaumont, V., Agrinier, P., Javoy, M., and Robert, F., 1994, Determination of the CO contribution to the $^{15}N/^{14}N$ ratio measured by mass spectrometry: Analytical Chemistry, v. 66, pp. 2187–9.

Beaumont, V. and Robert, F., 1999, Nitrogen isotope ratios of kerogens in Precambrian cherts: A record of the evolution of atmosphere chemistry?: Precambrian Research, v. 96, pp. 63–82.

Bebout, G. E. and Fogel, M. L., 1992, Nitrogen-isotope compositions of metasedimentary rocks in the Catalina Schist, California – implications for metamorphic devolatilization history: Geochimica et Cosmochimica Acta, v. 56, no. 7, pp. 2839–49.

Boyd, S. R. and Philippot, P., 1998, Precambrian ammonium biogeochemistry: a study of the Moine metasediments, Scotland: Chemical Geology, v. 144, pp. 257–68.

Brunner, B., Contreras, S., Lehmann, M. F., Matantseva, O., Rollog, M., Kalvelage, T., Klockgether, G., Lavik, G., Jetten, M. S. M., Kartal, B., and Kuypers, M. M. M., 2013, Nitrogen isotope effects induced by anammox bacteria: Proceedings of the National Academy of Sciences, v. 110, no. 47, pp. 18994–9.

Busigny, V., Lebeau, O., Ader, M., Krapez, B. , and Bekker, A., 2013, Nitrogen cycle in the Late Archean ferruginous ocean: Chemical Geology, v. 362, pp. 115–30.

De Pol-Holz, R., Robinson, R. S., Hebbeln, D., Sigman, D. M., and Ulloa, O., 2009, Controls on sedimentary nitrogen isotopes along the Chile margin: Deep-Sea Research Part Ii-Topical Studies in Oceanography, v. 56, no. 16, pp. 1100–12.

Freudenthal, T., Wagner, T., Wenzhofer, F., Zabel, M., and Wefer, G., 2001, Early diagenesis of organic matter from sediments of the eastern subtropical Atlantic: Evidence from stable nitrogen and carbon isotopes: Geochimica et Cosmochimica Acta, v. 65, no. 11, pp. 1795–808.

Fulton, J. M., Arthur, M. A., Thomas, B., and Freeman, K. H., 2018, Pigment carbon and nitrogen isotopic signatures in euxinic basins: Geobiology, v. 16, pp. 429–45.

Gao, X., Yang, Y., and Wang, C., 2012, Geochemistry of organic carbon and nitrogen in surface sediments of coastal Bohai Bay inferred from their ratios and stable isotopic signatures: Marine Pollution Bulletin, v. 64, pp. 1148–55.

Garvin, J., Buick, R., Anbar, A. D., Arnold, G. L., and Kaufman, A. J., 2009, Isotopic evidence for an aerobic nitrogen cycle in the latest Archean: Science, v. 323, pp. 1045–8.

Godfrey, L. V. and Falkowski, P. G., 2009, The cycling and redox state of nitrogen in the Archaean ocean: Nature Geoscience.

Granger, J., Sigman, D. M., Lehmann, M. F., and Tortell, P. D., 2008, Nitrogen and oxygen isotope fractionation during dissimilatory nitrate reduction by denitrifying bacteria: Limnology and Oceanography, v. 53, no. 6, pp. 2533–45.

Grasby, S., Beauchamp, B., Bond, D. P. G., Wignall, P. B., Talavera, C., Galloway, J. M., Piepjohn, K., Reinhardt, L., and Blomeier, D., 2015, Progressive environmental deterioration in northwestern Pangea leading to the latest Permian Extinction: GSA Bulletin, v. 127, pp. 1331–47.

Hoch, M. P., Fogel, M. L., and Kirchman, D. L., 1992, Isotope fractionation associated with ammonium uptake by a marine bacterium: Limnology and Oceanography, v. 37, no. 7, pp. 1447–59.

Ishida, A., Kitajima, K., Williford, K. H., Tuite, M. L., Kakegawa, T., and Valley, J. W., 2018, Simultaneous in situ analysis of carbon and nitrogen isotope ratios in organic matter by secondary ion mass spectrometry: Geostandards and Geoanalytical Research, pp. 1–15.

Junium, C. K. and Arthur, M. A., 2007, Nitrogen cycling during the cretaceous, Cenomanian-Turonian oceanic anoxic event II: Geochemistry Geophysics Geosystems, v. 8.

Kipp, M. A., Stueken, E. E., Yun, M., Bekker, A. , and Buick, R., 2018, Pervasive aerobic nitrogen cycling in the surface ocean across the Paleoproterozoic era: Earth and Planetary Science Letters, v. 500, pp. 117–26.

Knies, J., Grasby, S., Beauchamp, B., and Schubert, C. J., 2013, Water mass denitrification during the latest Permian extinction in the Sverdrup Basin, Arctic Canada: Geology, v. 41, pp. 167–70.

Koehler, M. C., Buick, R., Kipp, M. A., Stueken, E. E., and Zaloumis, J., 2018, Transient surface ocean oxygenation recorded in the ~2.66-Ga Jeerinah Formation, Australia: Proceedings of the National Academy of Sciences, v. 2018, pp. 1–6.

Kump, L. R., Junium, C. K., Arthur, M. A., Brasier, A., Fallick, A., Melezhik, V., Lepland, A., Crne, A. E., and Luo, G., 2011, Isotopic evidence for massive oxidation of organic matter following the Great Oxidation Event: Science, v. 334, pp. 1694–6.

Looy, C. V., Twitchett, R. J., Dilcher, D. L., Van Konijnenburg-Van Cittert, J. H. A., and Visscher, H., 2001, Life in the end-Permian dead zone: PNAS, v. 98, pp. 7879–83.

Macko, S. A. and Estep, M. F., 1984, Microbial alteration of stable nitrogen and carbon isotopic compositions of organic matter: Organic Geochemistry, v. 6, pp. 787–90.

Mandernack, K. W., Mills, C. T., Johnson, C. A., Rahn, T., and Kinney, C., 2009, The δ^{15}N and δ^{18}O values of N_2O produced during the co-oxidation of ammonia by methanotrophic bacteria: Chemical Geology, v. 267, pp. 96–107.

Megonigal, J.P., Hines, M.E., Visscher, P.T., 2003, 8.08 – Anaerobic Metabolism: Linkages to Trace Gases and Aerobic Processes. In: Treatise on Geochemistry, Holland and Turekian (eds.), pp. 317–424.

Mettam, C., Zerkle, A. L., Claire, M. C., Izon, G., Junium, C. K., and Twitchett, R. J., 2017, High-frequency fluctuations in redox conditions during the latest Permian mass extinction: Palaeogeography Palaeoclimatology Palaeoecology, v. 485, pp. 210–23.

Meyers, P. A., 1997, Organic geochemical proxies of paleoceanographic, paleolimnologic, and paleoclimatic processes: Organic Geochemistry, v. 27, pp. 213–50.

Möbius, J., Lahajnar, N., and Emeis, K. C., 2010, Diagenetic control of nitrogen isotope ratios in Holocene sapropels and recent sediments from the Eastern Mediterranean Sea: Biogeosciences Discussions, v. 7, pp. 1131–65.

Nishizawa, M., Miyazaki, J., Makabe, A., Koba, K., and Takai, K., 2014, Physiological and isotopic characteristics of nitrogen fixation by hyperthermophilic methanogens: Key insights into nitrogen anabolism of the microbial communities in Archean hydrothermal systems: Geochimica et Cosmochimica Acta, v. 138, pp. 117–35.

Olsen, S. L., Kump, L. R., and Kasting, J. F., 2013, Quantifying the areal extent and dissolved oxygen concentrations of Archean oxygen oases: Chemical Geology, v. 362, pp. 35–43.

Papineau, D., Mojzsis, S. J., Karhu, J. A., Marty, B., 2005, Nitrogen isotopic composition of ammoniated phyllosilicates: case studies from Precambrian metamorphosed sedimentary rocks: Chemical Geology, v. 216, pp. 37–58.

Peters, K. E., Sweeney, R. E., and Kaplan, I. R., 1978, Correlation of carbon and nitrogen stable isotope ratios in sedimentary organic matter: Limnology and Oceanography, v. 23, pp. 598–604.

Pinti, D. L. and Hashizume, K., 2011, Early Life Records from Nitrogen Isotopes. In: Golding, S., and Glikson, M. , eds., Earliest Life on Earth: Habitats, Environments, and Methods of Detection: Dordrecht, Springer.

Polissar, P. J., Fulton, J. M., Junium, C. K., Turich, C. H., and Freeman, K. H., 2009, Measurement of ^{13}C and ^{15}N isotopic composition on nanomolar quantitities of C and N: Analytical Chemistry, v. 81, pp. 755–63.

Robinson, R. S., Kienast, M., Albuquerque, A. L., Altabet, M., Contreras, S., De Pol Holz, R., Dubois, N., Francois, R., Galbraith, E., Hsu, T.-C., Ivanochko, T., Jaccard, S., Kao, S.-J., Kiefer, T., Kienast, S., Lehmann, M. F., Martinez, P., McCarthy, M., Moebius, J., Pedersen, T., Quan, T. M., Ryabenko, E., Schmittner, A., Schneider, R., Schneider-Mor, A., Shigemitsu, M., Sinclair, D., Somes, C., Studer, A., Thunell, R. , and Yang, J.-Y., 2012, A review of nitrogen isotopic alteration in marine sediments: Paleoceanography, v. 27.

Saitoh, M., Ueno, Y., Nishizawa, M., Isozaki, Y., Takai, K., Yao, J., and Ji, Z., 2014, Nitrogen isotope chemostratigraphy across the Permian-Triassic boundary at Chaotian, Sichuan, South China: Journal of Asian Earth Sciences, v. 93, pp. 113–28.

Schoepfer, S. D., Henderson, C. M., Garrison, G. H., and Ward, P. D., 2012, Cessation of a productive coastal upwelling system in the Panthalassic Ocean at the Permian-Triassic boundary: Palaeogeography Palaeoclimatology Palaeoecology, v. 313, pp. 181–8.

Stüeken, E. E., Buick, R., Guy, B., and Koehler, M. C., 2015a, Isotopic evidence for biological nitrogen fixation by Mo-nitrogenase from 3.2 Gyr: Nature, v. 520, p. 666–9.

Stüeken, E. E., Buick, R., and Schauer, A. J., 2015b, Nitrogen isotope evidence for alkaline lakes on late Archean continents: Earth and Planetary Science Letters, v. 411, p. 1–10.

Stüeken, E. E., Kipp, M. A., Koehler, M. C. , and Buick, R., 2016a, The evolution of Earth's biogeochemical nitrogen cycle: Earth-Science Reviews, v. 160, p. 220–39.

Stüeken, E. E., Zaloumis, J., Meixnerova, J., and Buick, R., 2017, Differential metamorphic effects on nitrogen isotopes in kerogen extracts and bulk rocks: Geochimica et Cosmochimica Acta, v. 217, p. 80–94.

Thomazo, C., Ader, M., and Philippot, P., 2011, Extreme 15 N-enrichments in 2.72-Gyr-old sediments: evidence for a turning point in the nitrogen cycle: Geobiology, v. 9, no. 2, p. 107–20.

Voss, M., Bange, H. W., Dippner, J. W., Middelburg, J. J., , Montoya, J. P., Ward, B., 2013, The marine nitrogen cycle: recent discoveries, uncertainties and the potential relevance of climate change. Philosophical Transactions Royal Society of London B Biological Science. v. 368.

Weiss, M. C., Sousa, F. L., N. , Neukirchen, S., Roettger, M., Nelson-Sathi, S., and Martine, W. F. , 2016, The physiology and habitat of the last universal common ancestor: Nature Microbiology, v. 1, p. 16116.

Yang, J., Junium, C. K., Grassineau, N. V., Nisbet, E. G., Izon, G., Mettam, C., Martin, A., and Zerkle, A. L., 2019, Ammonium availability in the Late Archaean nitrogen cycle: Nature Geoscience, v. 12, no. 7, p. 553–7.

Zerkle, A. L., House, C. H., Cox, R. P., and Canfield, D. E., 2006, Metal limitation of cyanobacterial N_2 fixation and implications for the Precambrian nitrogen cycle: Geobiology, v. 4, p. 285–97.

Zerkle, A. L., Poulton, S. W., Newton, R. J., Mettam, C., Claire, M. W., Bekker, A. , and Junium, C. K., 2017, Onset of the aerobic nitrogen cycle during the Great Oxidation Event: Nature, v. 542, p. 465–7.

Cambridge Elements

Elements in Geochemical Tracers in Earth System Science

About the Series

This innovative series provides authoritative, concise overviews of the many novel isotope and elemental systems that can be used as 'proxies' or 'geochemical tracers' to reconstruct past environments over thousands to millions to billions of years – from the evolving chemistry of the atmosphere and oceans to their cause-and-effect relationships with life.

Covering a wide variety of geochemical tracers, the series reviews each method in terms of the geochemical underpinnings, the promises and pitfalls, and the "state-of-the-art" and future prospects, providing a dynamic reference resource for graduate students, researchers and scientists in geochemistry, astrobiology, paleontology, paleoceanography and paleoclimatology.

The short, timely, broadly accessible papers provide much-needed primers for a wide audience – highlighting the cutting edge of both new and established proxies as applied to diverse questions about Earth system evolution over wide-ranging time scales.

Cambridge Elements

Elements in Geochemical Tracers in Earth System Science

Elements in the Series

The Uranium Isotope Paleoredox Proxy
Kimberly V. Lau, Stephen J. Romaniello and Feifei Zhang

Triple Oxygen Isotopes
Huiming Bao

Application of Thallium Isotopes
Jeremy D. Owens

Earth History of Oxygen and the iprOxy
Zunli Lu, Wanyi Lu, Rosalind E. M. Rickaby and Ellen Thomas

Selenium Isotope Paleobiogeochemistry
Eva E. Stüeken and Michael A. Kipp

The TEX_{86} Paleotemperature Proxy
Gordon N. Inglis and Jessica E. Tierney

The Pyrite Trace Element Paleo-Ocean Chemistry Proxy
Daniel D. Gregory

Calcium Isotopes
Elizabeth M. Griffith and Matthew S. Fantle

Pelagic Barite: Tracer of Ocean Productivity and a Recorder of Isotopic Compositions of Seawater S, O, Sr, Ca and Ba
Weiqi Yao, Elizabeth Griffith and Adina Paytan

Vanadium Isotopes: A Proxy for Ocean Oxygen Variations
Sune G. Nielsen

Cerium Anomalies and Paleoredox
Rosalie Tostevin

Nitrogen Isotopes in Deep Time
Colin Mettam and Aubrey L. Zerkle

A full series listing is available at: www.cambridge.org/EESS

Printed in the United States
By Bookmasters